Newton & Descartes's Night in Madrid

MATH MUSICALS

www.MathMusicals.com

Volume 1

by

Anne Lazo
and
Michael Wiskar

BIG IDEAS LEARNING®
Erie, Pennsylvania
BigIdeasLearning.com

Cartoon Illustrations: MoreFrames Animation
Photo of Michael Wiskar provided by Greg Blue *www.gregblue.com*

Stockphoto credits:
exxorian/DigitalVision Vectors/Getty Images
dimonspace/iStock/Getty Images Plus
mrhighsky/iStock/Getty Images Plus
Akrain/iiStock/Getty Images Plus
wissanu99/iStock/Getty Images Plus
Mjak/iStock/Getty Images Plus

Printed in the United States.

ISBN 13: 978-1-63598-920-5

123456789–22 21 20 19 18

Newton *Descartes*

Contents

The authors of *Big Ideas Math* understand the importance of storytelling and music in learning. Storytelling puts math in real-life contexts, and music helps children remember. In collaboration with storyteller Anne Lazo and musician Michael Wiskar, the authors of *Big Ideas Math* are excited to offer *Math Musicals*—engaging musicals that bring mathematics to life!

In this fictional world, students can interact with many different characters, including Newton and Descartes! Students will often wonder—What adventures will Newton and Descartes have today? What will they learn? What will they do next? Using *Math Musicals* will not only engage students in the mathematics, but will help improve their literacy skills as well.

Math and literacy skills are critically important in this world. Bringing math and literacy together through music is the perfect way to instill a lifelong love of learning, encouraging young minds to love math. At Big Ideas Learning, we believe providing the opportunity for students to interact with and learn the math through storytelling and music is just one more way to make learning visible.

About the Authors iv
Math Musicals Stories and Topics.......... v
Accompanying Products vii

Newton & Descartes's Night in Madrid

Tiny Tapas (Multiplying by 3).......... 1
Flamenco Mice (2s, 3s, 4s, and 5s Multiplication Facts).......... 15
Where is the Park? (Fractions) 27
Fútbol Fun (Perimeter).......... 41
Best Friends (I Get You and You Get Me).......... 56

Sheet Music

Tiny Tapas.......... 58
Olé, Olé, Olé! (Welcome to Spain) 59
Break It Down Into Parts.......... 61
Almost Home (The Perimeter Song) 63
Best Friends (I Get You and You Get Me).......... 65

About the Authors

Anne Lazo is an international performing arts educator, theatrical director, and author and composer of children's musical educational books.

Anne graduated from the University of Southern California and has over 20 years experience teaching and managing inspirational arts programs and productions for students, as well as professional actors, musicians, and dancers.

Anne worked in Japan for 15 years as an accomplished musical theater director, acting coach, choreographer, script writer, director of educational programs, and motivational speaker for arts advocacy. She has worked for Disney's World of English (World Family English), NHK Television and Radio, and owned her own production company, where she produced and directed musicals for schools and companies. In the United States, Anne was the Director of Education for the New West Symphony, where she developed and implemented many community outreach programs for all ages throughout the Los Angeles area.

Anne currently lives with her husband in Spain and is the Creator and Founder of My Child and Me - English©. Anne and her theatrical team of actors and teachers provide ESL educational shows, musical storybooks, and classes for young children.

Michael (Mike) Wiskar is a Leo-nominated composer, songwriter, independent music producer, vocalist, and multi-instrumentalist.

A graduate of the Music Program at MacEwan University as well as Langara's Digital Music Production Program, Mike began his music career in his teens as a performer in rock, country, and jazz bands before working as a songwriter for various artists in Quebec. There he worked closely with Myles Goodwyn of legendary Canadian rock band April Wine, among other top-level songwriters and producers.

Mike has since written and produced songs and music scores for theater, short films, full length and series documentaries, choirs, contemporary artists, and numerous television series (Operation Vacation, Psycho Kitty, Men's Fashion Insider, Kitty 911, and Favorite Places, to name just a few).

Math Musicals Stories and Topics

Grade K

Newton & Descartes's Coolest, Rockin' Day Ever

- *Fish Crackers* (Counting to 10)
- *Cool Cats & Rockin' Dogs* (Partner Numbers to 5)
- *Big Bulldog Bob* (Counting to 20)
- *JAM Session* (Naming Two- and Three-Dimensional Shapes)
- *Best Friends (I Get You and You Get Me)*

JAM Session

100 Waves

Grade 1

Newton & Descartes's Day at the Beach

- *Seashells* (Addition to 10)
- *Cora's Home* (Subtraction to 10)
- *100 Waves* (Counting to 100 by 10s)
- *Racing the Clock* (Telling Time)
- *Best Friends (I Get You and You Get Me)*

Grade 2

Newton & Descartes's Four-Legged Fun

- *Four-Legged Games* (Two-Digit Addition)
- *Newton & Descartes's Turn* (Two-Digit Addition and Subtraction)
- *Pet Airlines* (Counting to 1,000 by 100s)
- *At the Airport* (Money)
- *Best Friends (I Get You and You Get Me)*

At the Airport

Grade 3

Newton & Descartes's Night in Madrid

- *Tiny Tapas* (Multiplying by 3)
- *Flamenco Mice* (2s, 3s, 4s, and 5s Multiplication Facts)
- *Where is the Park?* (Fractions)
- *Fútbol Fun* (Perimeter)
- *Best Friends (I Get You and You Get Me)*

Fútbol Fun

Newton Leads the Way

Grade 4

Newton & Descartes's Skate Park Adventure

- *Jump Start* (Estimating Sums and Products)
- *The Amphibian* (Divisibility Rules for 2, 3, and 5)
- *The Catwalk* (Adding and Subtracting Fractions with Like Denominators)
- *Newton Leads the Way* (Types of Measurement)
- *Best Friends (I Get You and You Get Me)*

Grade 5

Newton & Descartes's Pet Center Adventure

- *Happy Paws Club* (The Language of Decimals)
- *The Cat's Eye Club* (Multiplying Decimals and Whole Numbers)
- *Bark Like a Dog* (Adding Fractions with Unlike Denominators)
- *Friends* (Volume of a Rectangular Prism)
- *Best Friends (I Get You and You Get Me)*

Friends

Accompanying Products

Visit *www.MathMusicals.com* for the following products.

Songs

Math Musicals songs are available for use throughout the year. Play the songs from *Math Musicals* to not only help students learn concepts, but to help them review all year long as well!

Animated Videos

The *Math Musicals* songs and storybooks come to life in these engaging animations filled with vivid and fun images that your students are sure to enjoy. Join Newton and Descartes on adventures around the world! From rocking out at a jam party to flying high in an airplane, Newton and Descartes engage students in mathematics in fun and innovative ways!

Sheet Music

Each *Math Musical* song comes with its own sheet music. Students and teachers can engage in learning by playing along with songs from *Math Musicals* on a piano or a guitar! Teachers have the ability to truly mix the teaching of math and music with this easy-to-read sheet music.

Differentiated Rich Math Tasks

Differentiated Rich Math Tasks are a fun and intriguing way for students to interact with mathematics. These leveled worksheets challenge students to stretch their knowledge and dig deeper! Each task has three different levels so teachers can easily differentiate to meet the needs of every student.

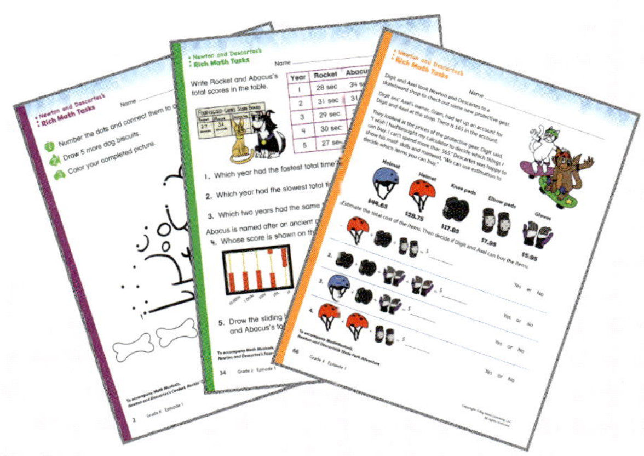

Puppets and Plush Toys

You can use the Newton and Descartes puppets and plush toys to make learning fun and to help students talk about math!

Newton & Descartes's Night in Madrid

by
Anne Lazo and Michael Wiskar

Tiny Tapas (Multiplying by 3)
Flamenco Mice (2s, 3s, 4s, and 5s Multiplication Facts)
Where is the Park? (Fractions)
Fútbol Fun (Perimeter)

TINY TAPAS

by Anne Lazo

Newton and Descartes were exhausted after flying for hours from America to Madrid, Spain. To make matters worse, they missed their bus to the center of the city. They had no idea that winning the Four-Legged Games would make them so famous! After spotting them at the airport, their new Spanish fans celebrated their victory into the night. When things finally settled down, Newton and Descartes found themselves alone, outside the airport.

"This is crazy!" cried Descartes. "Now what are we going to do? We missed our bus and only have a few euros left. We can't spend most of our money on another bus ticket!"

"Let's take the metro," yelped Newton. "I don't think it stops running for another few hours."

As they made their way to the metro station, they bumped into a mouse.

“Hola chicos!” said the mouse. “I recognize you from the televised Four-Legged Games in America. You are Newton and Descartes, the winners, right? What are you doing walking outside the airport so late?”

“We are trying to get to the center of the city to the Park Hotel,” meowed Descartes. “We missed our bus because so many pets recognized us and we couldn’t get away.”

"Pets in Spain love sports and were excited to see their heroes from America in person! Allow me to take you. My name is Pérez Mouse, or Ratón Pérez. I am also famous here in Spain. I travel all over the country, going from house to house, making sure that all of the children who lost a tooth today receive a gift."

"Wow, you mean like the tooth fairy?" asked Newton.

"Sí, exactly like the tooth fairy. In fact, the tooth fairy and I are good friends. We tell each other interesting stories about our adventures visiting so many children! Now, vamonos! Let's go before it gets too late. Come, let me show you the way!"

Ratón Pérez jumped onto Newton's doggie bag and rode the metro with them. After riding a few stops, he exclaimed that they should get off at the next stop.

"Is this where the hotel is?" asked Newton.

"It's on the way. Don't worry, we will get there. You must see some wonderful parts of the city first. Madrid looks glorious at night and is a lot of fun!" squeaked Pérez with delight.

"But, we really should go directly to the hotel," pleaded Newton. "We want to surprise Miss Polly, our owner."

"Don't worry, since she isn't expecting you, it won't matter what time you surprise her! Besides, I am the best tour guide in Madrid! I have to visit more houses tonight, so I won't keep you too long!" Pérez assured them.

Ratón Pérez guided them to the street level and toward his favorite tapas restaurant, Tiny Tapas. Newton and Descartes had to stoop down and squeeze through the door because it was so small. Clearly, this restaurant was made for mice!

“I bet you are hungry, no? You have to try Spain’s most famous food, tapas. This place has the best tapas in Madrid and is where I eat with my friends almost every day!” Pérez boasted.

Newton and Descartes had to navigate their way through the noisy crowd of mice in the restaurant. They could see that all of the mice were enjoying their small dishes of different kinds of food. It looked like the most popular food here was cheese.

As Ratón Pérez started to order, he said that, since the tapas were different sizes, he suggested that **they each order eight** cheesy olives, **seven** cheesy puntillitas, **six** cheesy cheese tapas, **five** cheesy croquetas, **four** cheesy jamón, **three** cheesy gambas, **two** Manchego cheese with crackers, and to finish it off, **one** famous Spanish tortilla on cheesy bread!

3 × 1

3 × 2

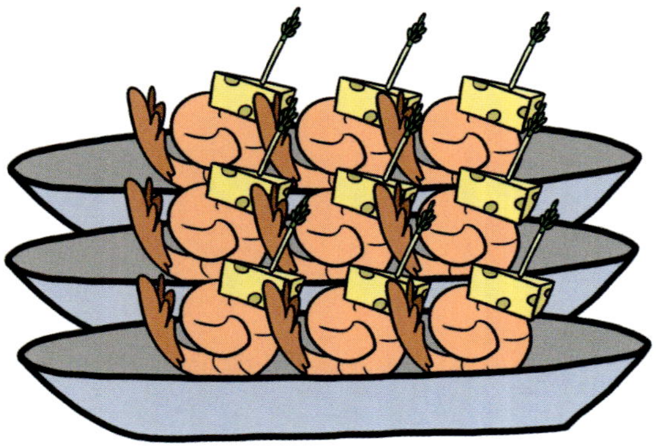

3 × 3

3 × 4

I can't eat all of that cheese. I am lactose intolerant!
That's 36 tapas each!
They're tiny! Mouse size.

3×5

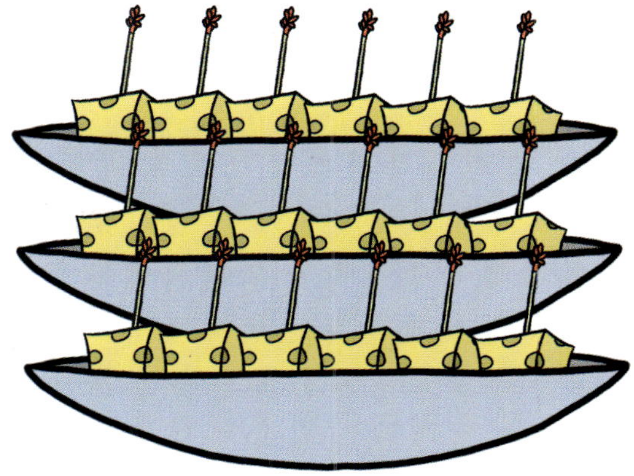

3×6

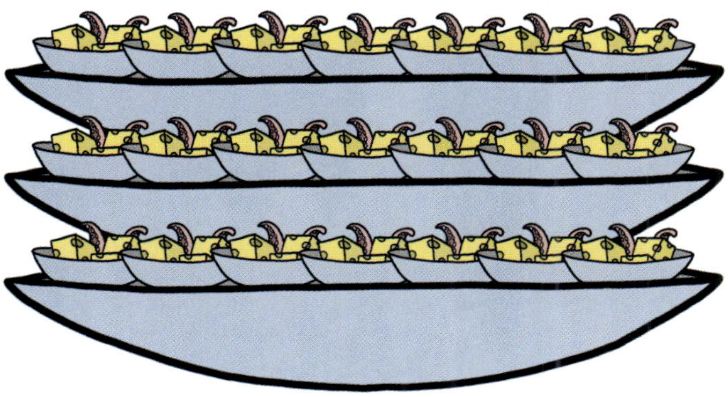

3×7

3×8

"Oh no, that's a lot of cheese," thought Descartes nervously.

$3 \times 1 = 3$ Spanish tortillas on cheesy bread

$3 \times 2 = 6$ Manchego cheese with crackers

$3 \times 3 = 9$ cheesy gambas

$3 \times 4 = 12$ cheesy jamón

$3 \times 5 = 15$ cheesy croquetas

$3 \times 6 = 18$ cheesy cheese tapas

$3 \times 7 = 21$ cheesy puntillitas

$3 \times 8 = 24$ cheesy olives

As the food came to the table, Descartes looked at the pile of cheesy dishes and felt ill. "Newton," Descartes whispered, "I can't eat all of that cheese. I am lactose intolerant!"

Tiny Tapas

by Michael Wiskar

Manchego, Cabrales, Idiazabal. *El Mejor!*
Zamorano, Valdeón, Majorero. *Tiny Tapas!*

3 × 1 is 3 Spanish tortilla on cheesy bread.

3 × 2 is 6 Manchego cheese with crackers.

3 × 3 is 9 cheesy gambas. (What's gambas?)

3 × 4 is 12 cheesy jamón. Jamón!

Manchego, Cabrales, Idiazabal. *El Mejor!*
Zamorano, Valdeón, Majorero. *Tiny Tapas!*

3 × 5 is 15 cheesy croquetas.

3 × 6 is 18 cheesey cheese tapas.

3 × 7 is 21 cheesy puntillitas.

3 × 8 is 24 cheesy olives!

Manchego, Cabrales, Idiazabal. *El Mejor!*
Zamorano, Valdeón, Majorero. *Tiny Tapas!*

FLAMENCO MICE

by Anne Lazo

Newton ate most of the cheesy tapas so that Descartes would not get sick. Ratón Pérez could see that Descartes was not eating very much.

“Aren’t you hungry, my friend?” asked Pérez.

“I ate a lot on the airplane and I am not used to eating so late, but I tried the tapas and they are very good! Thank you for introducing them to us!” replied Descartes in a friendly tone.

I ate a lot on the airplane and I am not used to eating so late.
Aren't you hungry, my friend?

As Newton and Descartes tried to say goodbye and leave, Ratón Pérez jumped up on his chair and yelled out to all of his mice friends, "I think Newton likes tapas!" Pérez laughed as he motioned to Newton, who was feeling very full from eating so much cheese.

All of the mice laughed and cheered, "Hip, hip, hooray! Here's to our new amigos who love our cheesy tapas!"

There were some flamenco mice who were not cheering. They hid in the corner of the room, under a table. They were clearly afraid of Descartes and did not understand why a cat was allowed in the restaurant. They did not realize that Newton and Descartes were famous and visiting from America. Pérez quickly explained who Newton and Descartes were, and that it was safe to come out.

"Don't worry, amigos, this cat is a friendly cat!" Pérez explained to the flamenco mice.

"Señor Descartes won't hurt you! He and Señor Newton are my friends! Why don't you play some music for them and dance? They will love our Spanish music!"

The flamenco mice came out from under the table. They figured that if Newton and Descartes were friends of Ratón Pérez, then they were safe.

The flamenco mice were happy to perform for them. They started to clap, strum their guitars, and dance. All of the mice in the room gathered around and joined in the song by counting, clapping, and cheering.

"Your feet are moving so fast. I am trying to count how many steps you take in 10 seconds," meowed Descartes.

"Instead of counting by ones, it's easier if you count by a larger number," said the flamenco dancer. "Here, **try counting the steps in this dance by twos**."

Descartes tried it. "**2, 4, 6, 8, 10, 12, 14, 16, 18, 20, 22, 24, 26, 28, 30.** I did it," Descartes said proudly. "You took 30 steps in 10 seconds."

"Now **try counting by threes**," said the dancer.

Descartes tried again. "**3, 6, 9, 12, 15, 18, 21, 24, 27, 30**. Wow, **counting by threes** was even easier than **counting by twos**!"

"Hey, Newton, you try it. See if you can count how many steps the dancers take in 20 seconds," challenged Descartes. "Here lay low to the floor and it will be easier. Try **counting by fives**."

Newton tried it. He layed flat on the floor and the dancers started a new dance. "**5, 10, 15, 20, 25, 30, 35, 40, 45, 50**," barked Newton. "That was so easy. I like **counting by fives**."

The flamenco dancers were happy to realize that Newton and Descartes were friendly and fun to be with. They all got together and sang a song to their new American friends.

Olé! Olé! Olé! (Welcome to Spain)

by Michael Wiskar

2 – 4 – 6 – 8
Our American friends are really great!
We go **10 – 12 – 14 – 16**
We've had our tapas now
let's start dancing.
3 – 6 – 9 – 12
We really know how to enjoy ourselves.
We go **15 – 18 – 21 – 24**
Now its time to hit the dance floor!

Olé, Olé, Olé!
Welcome to Spain.
Let the music play
Until the break of day.

Olé, Olé, Olé!
Welcome to Spain.
We'd love it if you'd stay
And dance the night away.
Olé, Olé, Olé!

4 – 8 – 12 – 16
You my friends, are pretty nifty.
20 – 24 – 28 – 32
Now my friends, here's what
we're going to do
5 – 10 – 15 – 20
You are in a place of plenty.
25 – 30 – 35 – 40
Now its time to start the party!

Olé, Olé, Olé!
Welcome to Spain.
Let the music play
Until the break of day.

Olé, Olé, Olé!
Welcome to Spain.
We'd love it if you'd stay
And dance the night away.
Olé, Olé, Olé!

"Look, my friends, time has gone by so quickly! I am very sorry to leave you now, but I really must go and visit more children who have lost their teeth. I must leave a gift for them before they wake up. But, don't worry, amigos, your hotel is not too far from here. Just follow this street all the way down until you reach the park. Your hotel is directly in front of the park. You can't miss it! I really must go now!" Ratón Pérez said as he made his way to the door.

"Hasta luego, amigos! I hope to see you again soon!" Ratón Pérez dashed out of the door and disappeared into the night.

WHERE IS THE PARK?

by Anne Lazo

Newton tried to fit through the door of the tapas restaurant to leave, but he had eaten too many tapas and could not get through.

"Look at that, he ate too many tapas!" squeaked one of the mice. All of the mice had to push Newton until he could squeeze out.

"Uno, dos, tres, PUSH!" **One fourth** of Newton was out.
"Uno, dos, tres!" **One half** of Newton was out.
"Uno, dos, tres, PUSH!" **Three fourths** of Newton was out.
"Uno, dos, tres!" Newton was finally the **whole** way out of the door!

"Yay, we did it! We are happy you enjoyed our food and music! Please come back again, our American friends!" exclaimed the joyous mice as they bid them farewell!

Newton and Descartes looked at each other and smiled. They could not believe what a strange and wonderful time they were having in this beautiful city! As they walked down the city streets, they passed old, fancy, tall buildings. Some of the buildings had statues of animals on or next to them. They saw regal horses, birds, and majestic lions that looked as though they were staring right at you as you walked by.

"Wow, the animals look frozen!" barked Newton. "I wonder if there are any dog or cat statues!"

"That would be weird!" meowed Descartes. "But, you never know, there are big cats, so maybe there are small cats too!" He and Newton laughed as they jumped up on a big, bronze lion statue. From the top of the lion they looked all around. The city lights were twinkling and the streets were bustling with action. There were so many owners and pets walking around and lots of commotion coming from inside the restaurants that lined the city streets.

They could not believe how busy Madrid was at night! Newton and Descartes jumped off of the statue and started to walk around the city again, looking for the park. Descartes wished he had brought a map with them. Ratón Pérez told them to walk straight and they would eventually make it to the park. Walk straight? Walk straight? The streets were starting to curve and he felt a bit nervous.

"I guess we got too excited looking around and got off track, Newton. Which way should we go now?" Descartes meowed nervously. "We didn't bring a map!"

"Don't worry, Descartes. I can find the park. I can find anything!" Newton replied as he sniffed around for clues. Newton was hoping to find grass, flower, or water scents that he could follow to the park.

"We can't sniff our way to the park, Newton," said Descartes. "Look, there is a parrot in the plaza sitting on a windowsill outside that restaurant. Let's ask him which way to go."

Descartes could not believe that even the parrot was awake at this late hour as he rushed over to ask him for directions. "Hello. Could you please tell us where the park is? We are looking for the Park Hotel. Do you know where it is?" The parrot just looked at him and did not answer.

"Hello. Hello! Please, can you tell me where the park is?" Descartes asked again in a loud voice. Again, the parrot just looked at him and did not answer. Descartes was starting to get frustrated, wondering why this parrot was being so rude.

"Hello, my name is Descartes and this is Newton. Unfortunately, we forgot to bring a map and now we are lost. Could you please tell us how to get to the park from here? Hello? Hello! Hello!"

Newton had to stop Descartes before he got too angry at this poor parrot. "Descartes," Newton whispered, "he doesn't speak English. You have to say, 'hola'!"

"Oh, of course, silly me!" Descartes chuckled and asked for directions again. "Hola. ¿Dónde está el parque? Where is the park?" Descartes asked again.

"Hola… hola… hola," answered the parrot.

"Hola, the park? El parque?" Descartes asked again.

"Hola, the park? El parque?" the parrot repeated.

"Oh, we are not getting anywhere," Descartes meowed as he shook his head in defeat.

"Oh, we are not getting anywhere. El parque. El parque," repeated the parrot again.

"Hello, may I help you? I am sorry, but I couldn't help but overhear you trying to speak with this parrot," said a friendly tourist cat. He was wearing a funny hat and a camera around his neck. "I just finished dinner and now I am on my way back to my hotel. Are you lost?"

"Are you from America too? Wow, you are eating so late at night!" Newton said, surprised.

"I'm from England and I've been touring Spain for 2 weeks. Everyone eats very late at night in Spain," said the tourist cat.

"Wow, that's crazy!" barked Newton. "If I have to eat so late every night, I'm going to starve!"

"Oh, you get used to it. They eat lunch later too, so it's not a big deal. Just think of each waking day as having 18 hours. You get to eat every 3 hours!" explained the tourist cat.

Begin Day:	Desayuno (Breakfast):	**8:00 A.M.**
One sixth:	Merienda a media mañana (Midmorning snack):	**11:00 A.M.**
One third:	Almuerzo (Lunch):	**2:00 P.M.**
One half:	Merienda media tarde (Afternoon snack):	**5:00 P.M.**
Two thirds:	La hora el aperitivo (Tapas Hour):	**8:00 P.M.**
Five sixths:	La Cena (Dinner):	**11:00 P.M.**

"Now, where are you trying to go? I have a map that I don't need anymore. Would you like to have it?" asked the tourist cat.

"That would be great. Thank you!" Descartes meowed, delighted to be able to see where they are. "We are looking for the Park Hotel, which is supposed to be in front of the big park. Do you know where the Park Hotel is?"

"I don't know that hotel, but if it's near the big park called Retiro Park, then I can show you were that is," replied the cat.

The tourist cat showed Newton and Descartes the map of the city. He explained that, now, they are approximately **a quarter of the way** to Retiro Park. They have to walk down the street to the big road. There, they will see a giant Spanish flag waving in the air. At that point, they will be **half way** to their destination. They must turn right at the flag and continue straight until they come upon a huge plaza with a giant fountain and a statue of a Roman goddess being drawn by two lions. At that point, they will be **three fourths of the way** to the park. Next, they should turn left and follow that road until they see a giant archway, La Puerta de Alcalá. Retiro Park is just behind the archway.

It sounded easy and not too far!

"Thank you so much!" meowed Descartes. "It's very kind of you to help us!"

"Well, I know what it's like to get here and not know where you are going. Have a great time. It's a wonderful city!" said the tourist cat as he waved goodbye and continued walking towards his hotel in the other direction. Newton and Descartes felt excited again. Now they knew where to go and were determined to not get distracted this time! As they started to walk towards the big Spanish flag, they could hear a screechy voice calling out in the background behind them.

"Hola, where's the park? Where's the park? ¿Dónde está el parque? ¿Dónde está el parque?"

Break It Down Into Parts

by Michael Wiskar

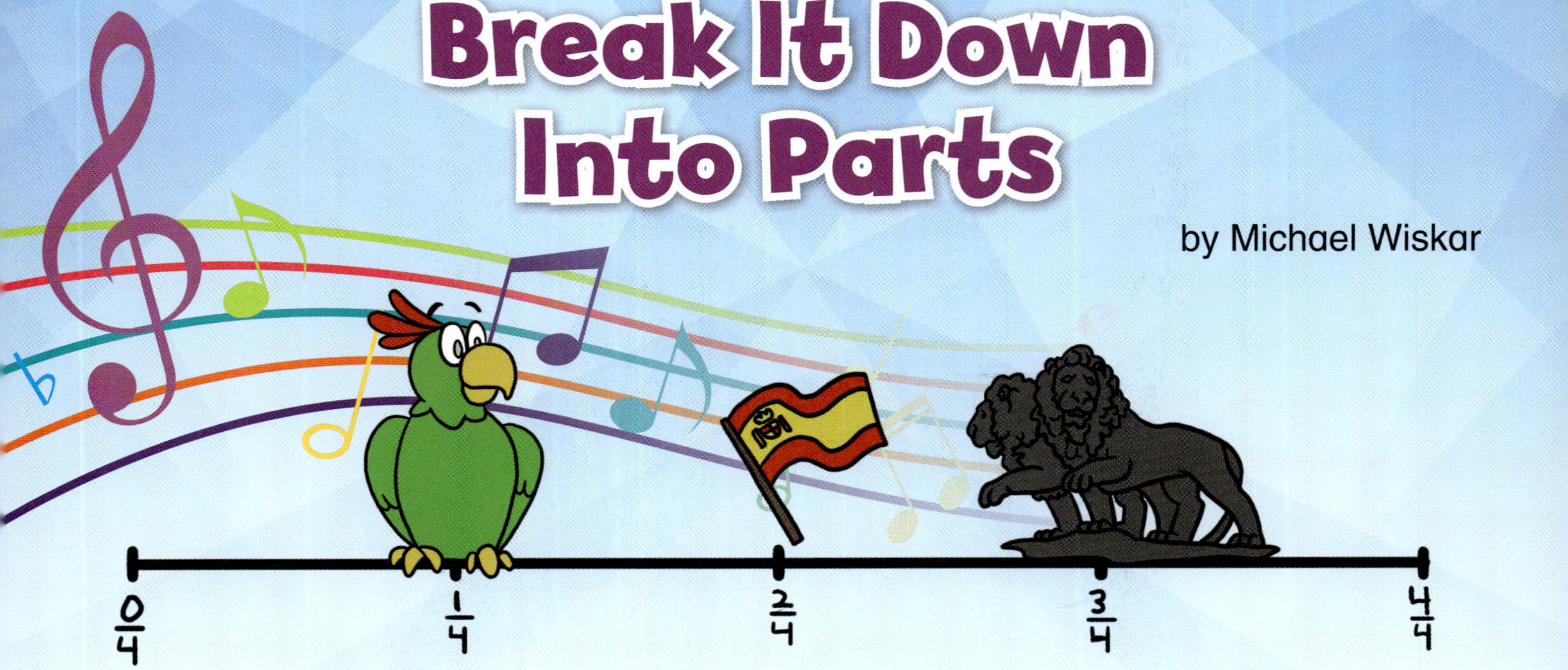

When you've got so far to go,
Here's something you've got to know.
You've got to break it down,
break it down,
You've got to break it down into parts.
If your trip is really long
Just listen to this silly song
And break it down, break it down,
You've got to break it down into parts.

So you've got to get to the park.
It's kind of dark
And it seems really far.
So, you break it down into parts
Because you're smart
And you'll get a good start.

First part: (What do you see?)
You get to the parrot
And you're **a quarter** of the way there.
Second part: (What do you see?)
It's a great big flag and you're **halfway** there now!

When you've got so far to go,
Here's something you've got to know.
You've got to break it down,
break it down,
You've got to break it down into parts.
If your trip is really long
Just listen to this silly song
And break it down, break it down
You've got to break it down into parts.

Third part: (What do you see?)
Two stone lions
Three quarters of the way now.
Fourth part: (What do you see?)
You're at the park, yeah you made
it **all the way** now! YEAH!

When you've got so far to go,
Here's something you've got to know.
You've got to break it down,
break it down,
You've got to break it down into parts.
If your trip is really long
Just listen to this silly song
And break it down, break it down,
You've got to break it down into parts.

Break it down into parts!

FÚTBOL FUN

by Anne Lazo

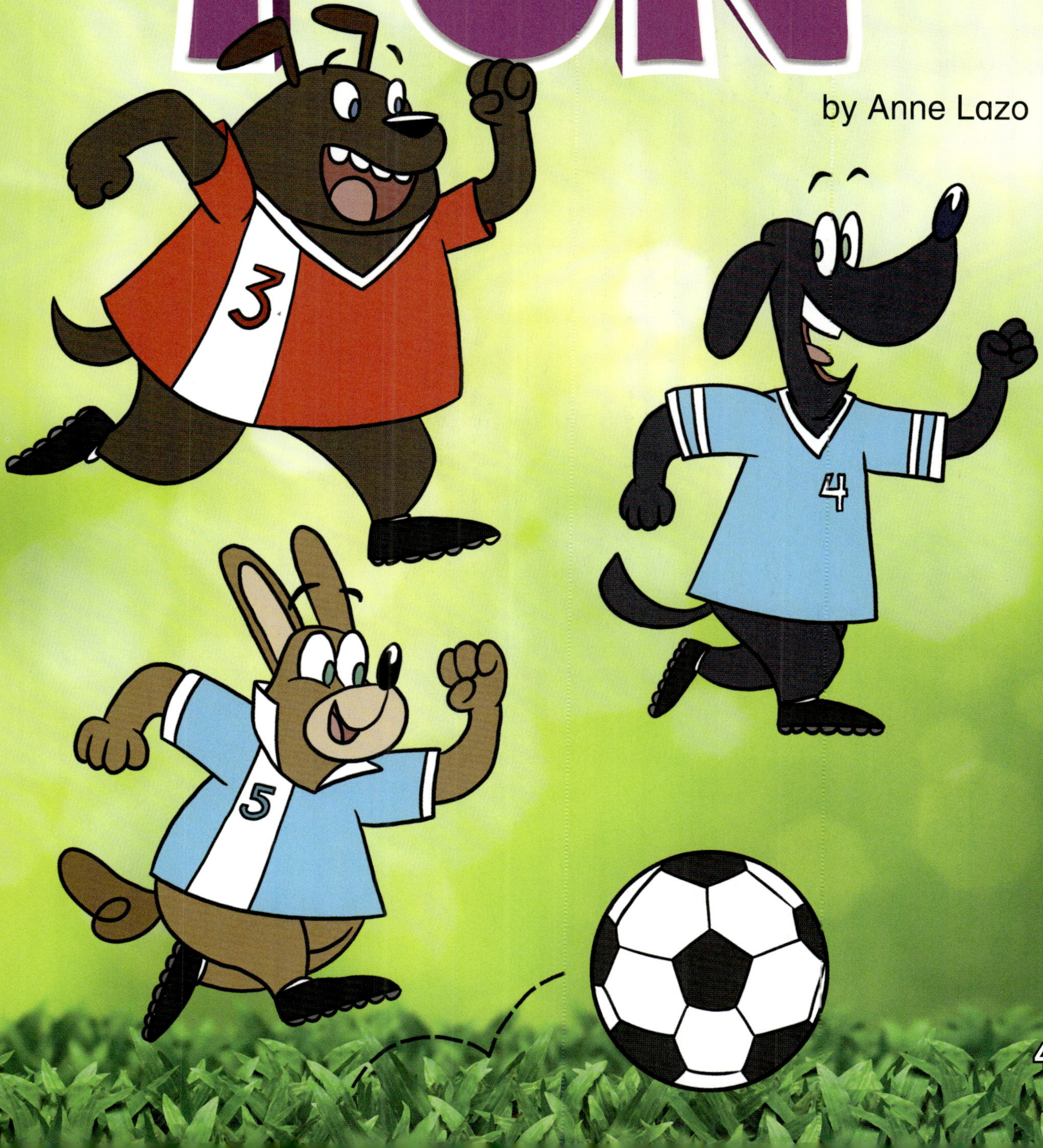

Retiro Park was closed by the time Newton and Descartes found it, but they were thrilled that they made it going step by step, following the map. Ratón Pérez told them that the Park Hotel was in front of the park, but the park was so big. They had no idea where to start looking.

"Let's ask someone where the hotel is," meowed Descartes.

"How big do you think the park is?" asked Newton. "Maybe we can just start **walking around the perimeter** and look for the hotel on the way."

"Yes, it must be somewhere close by," Descartes replied. "We must be close!"

Newton and Descartes were both exhausted, but they continued to **walk along the perimeter of the park**, dragging their luggage behind them. They could hear some raucous play and shouting in the distance and wondered what was going on. Before they knew it, a soccer ball went flying in the air and hit Newton square on the head. Newton jumped with fright, which caused the ball to fly over the fence and into the park.

"Newton, are you alright?" gasped Descartes. "That was quite a hit!"

"What was that?" Newton stammered, as he tried to get back up on his feet.

"I think it was a soccer ball. Those dogs over there must have been playing w th it and kicked it over here," explained Descartes.

Three dogs ran over to Newton and apologized. "¡Perdón, perdón, amigo!" barked one of the soccer dogs. "¡Lo siento!"

"No hablo español bien. I don't speak Spanish very well, but I am okay, don't worry," Newton answered, speaking slowly.

"That was a great header!" the younger dog, who spoke English very well, replied. "That ball went flying over the fence! Where are you from? Do you play fútbol in your country?"

"Fútbol? You mean soccer?" Newton answered.

"Amigo, in Spain we call the game fútbol!" the young dog continued. "You are a natural! Hey, wait a minute, aren't you guys the winners of the Four-Legged Games?"

"Yes, but…," replied Newton.

"No wonder the ball went flying! You're amazing!" chuckled the young dog.

Descartes thought it was ridiculous that, now, every pet in the world thinks that he and Newton are great athletes. He just rolled his eyes as the fútbol dogs all started laughing and giving each other and Newton high-fives.

"Sorry about losing your ball. We can get your ball back in the morning when the park re-opens. We are staying in a hotel near here, so we can come out early and look for it," Descartes interrupted impatiently.

"We can get it now Descartes!" barked Newton. "You can climb the tree and jump over to the other side of the fence and get the ball."

"Me? And how am I supposed to carry the ball out?" meowed Descartes, irritated.

"Just check the situation out and let us know if you can see a way out. Besides, from up in the tree you will be able to see if the hotel is on this side of the park," suggested Newton. Descartes agreed to try, but only because Newton had the good idea of checking for the hotel from a high vantage point. He scurried up a tree and had a good look around. Descartes saw some hotels across the street, but none of them said "Park Hotel."

Descartes leapt down off of the tree and searched the ground for the ball. It took him awhile to find it because the park was so dark. He eventually found it, not on the ground, but stuck in another tree.

"I found it!" Descartes meowed loudly. "But, it's stuck in a tree. I'll try to get it back over the fence. Get ready to catch it!" Descartes knew that he probably could not pick the ball up and throw it hard enough to make it over the fence, so he put his thinking cap on and thought of another way. "I'll just pull the tree branch way back and use it like a sling shot to get the ball over to them," he thought.

The ball went flying into the sky, like Descartes thought, but it bounced off of the fence and fell back into the park.

"It didn't make it over the fence. Now it's on the ground," Descartes yelled out to the fútbol dogs.

Just then, a park worker, close to Newton, opened a gate slightly to take out some trash. Newton grabbed the luggage and, thinking quickly, ran toward the gate with the fútbol dogs. They tiptoed past the worker and into the park before the worker closed the gate again. Luckily for them, the worker did not see them because he was carrying a big trash can that blocked his view.

"How did you get in?" Descartes asked.

"A worker opened the gate and we rushed in before he closed it," answered Newton.

"So, how are we, I mean all of you, going to get out now?" Descartes replied as he pictured himself getting out the same way he came in and leaving them all behind.

"Let's stay in the park for the rest of the night and play fútbol!" barked the dogs. Descartes was so tired that all he wanted to do was sleep! He could not believe what they had been through already, in just one night in Madrid. "Why were these dogs playing fútbol so late, and why was the whole city so busy all night?" he thought. "Shouldn't everyone be sleeping?" He and Newton were experiencing so many cultural differences. But as he looked at the fútbol dogs, he could see how happy and playful they were, not a care in the world. They loved fútbol and it did not matter what time of day they played, they just wanted to play!

The fútbol dogs made a makeshift playing field that was lit, just enough, by some lamp posts. They dribbled the ball back and forth, back and forth. Newton joined in while Descartes laid down and watched. "This I have to see," he thought. "I don't think Newton knows how to play fútbol! But then again, Newton surprises me all of the time. Whatever he decides he is going to do, he gives it his all and has fun."

"Go, Newton, go!" Descartes yelled out." If you can't beat them, join them," Descartes surrendered.

Even though Newton was tired, he tried his best. When the ball was finally passed to him, he went flying down the wing, cut inside the center, lined up the ball, and shot the ball straight into the goal. He scores! GOAAAL!

All of the fútbol dogs jumped up and down, barking with excitement! Even Descartes stood up and got caught up in the thrill of the game, well, kind of a game. They were just playing around, when suddenly, the exhilaration of playing the sport with their new friends made them forget about how tired they were.

For the moment, Newton and Descartes were able to forget about all of the obstacles they had been facing, and just enjoy their new experiences. Perhaps, in the morning they could try to look for the hotel again. It would be easier in the daylight anyway.

Newton, Descartes, and the fútbol dogs fell asleep in the park and woke up as the sun shined on their faces. The gates of Retiro Park re-opened and the morning hustle and bustle of the city began. The fútbol dogs explained to Newton and Descartes that **the perimeter of the park was only three miles long**. It was **one mile on two sides** and **one half of a mile on the other two sides**.

$$\textbf{Perimeter} = 1 + \frac{1}{2} + 1 + \frac{1}{2} = 3 \textbf{ miles}$$

It would not be hard to walk around the entire park to find the hotel. It must be across the street from one of the sides!

Newton and Descartes thanked their new friends and parted ways. They both felt rejuvenated and happy. With suitcases in hand, they continued their journey around the perimeter of the park.

Almost Home (The Perimeter Song)

by Michael Wiskar

We've seen a lot here you and me.
We've made some friends that feel like family.
We ate too much and went dancing.
Didn't we?

And now this last part of our journey
Around the park to find Polly.
There's nowhere else I would rather be,
Just you and me.

We go traveling around the **perimeter**.
Up one side, across the other
Around and down the **perimeter**.
One more side and then we're almost home.

(Descartes) "Well, not *home* exactly."
(Newton) "Wherever Polly is, is home."
(Descartes) "Yeah, you're right."

We go traveling around the **perimeter**.
Up one side, across the other.
Around and down the **perimeter**.
One more side and then we're almost home.

Then we're almost home.
And then we're almost home.
And then we're almost home.
Then we're almost home.

(Descartes) "Wow, what an adventure."

(Newton) "I cannot wait to tell Polly
about everything we've seen and done!"

(Descartes) "Ummm, I don't think Polly
speaks dog, Newton."

(Newton) "Yeah, I don't think she speaks
cat either."

(Descartes) "Hmmm. So how come you and
I can understand each other?"

(Newton) "I don't know. I think good friends
just do. Even when they seem so different."

(Descartes) "Yeah."

(Newton) "Hey, I think I see her. I think
I see Polly!"

There it was, El Hostel del Parque. It was no wonder they could not find it on the map, they were not thinking in Spanish. That had to be it! Newton and Descartes were confident as they walked into the hotel. Miss Polly would never know all that they had been through to find her, but that did not matter. What mattered was that they finally made it, together!

There she was, having breakfast in the hotel restaurant. With their winner's collars on, they proudly ran over to her table.

Best Friends (I Get You and You Get Me)

by Michael Wiskar

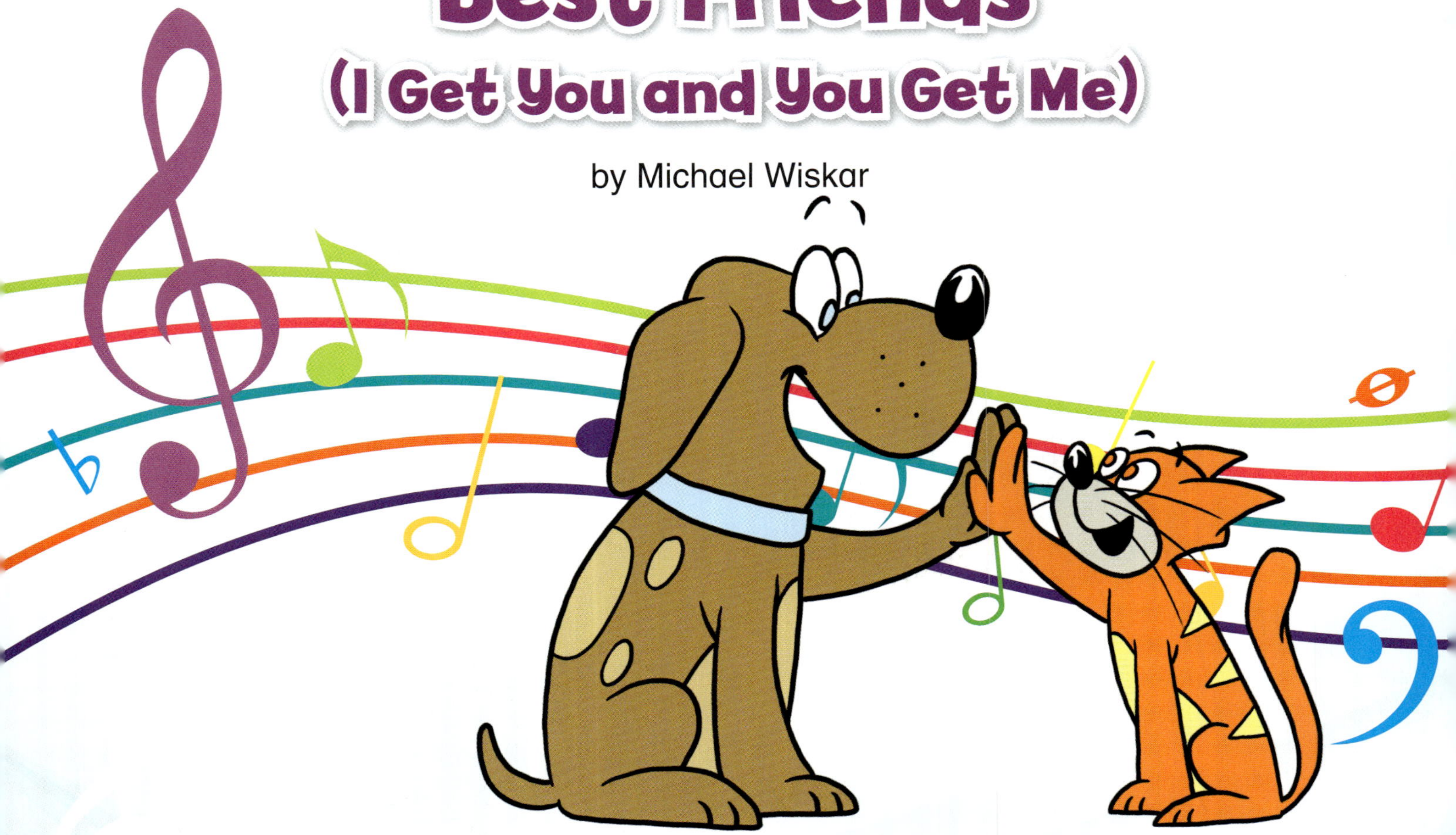

A wink, a nod, a sideways glance,
  a little elbow to the ribs.
We speak a secret language and
  no one else knows what it is.
You know me better than
  I even know myself.
It's a mutual connection
  between us and no one else.

You've got a friend,
Whenever you need a friend.
You've got a friend,
Whenever you need a friend.
I can't pretend
  there's anywhere else I'd rather be,
'Cause I get you and you get me.

And even when there's nothing to do
  we're never feeling bored.
We've had a thousand adventures
  and we'll have a thousand more.
You make me laugh so hard
  I feel like I could cry.
And we pick each other up
  when we're together,
We don't even have to try.

You've got a friend,
Whenever you need a friend.
You've got a friend,
Whenever you need a friend.
I can't pretend
  there's anywhere else I'd rather be,
'Cause you've got a friend in me.

'Cause I get you and you get me.
'Cause I get you and you get me.
We're a special kind of family.

And even if there's trouble brewing
we always know just what we're doing,
'Cause I've always got your back
and you've got mine.
And everything's fine!

You've got a friend,
Whenever you need a friend.
You've got a friend,
Whenever you need a friend.
I can't pretend
there's anywhere else I'd rather be,
'Cause I get you and you get me.

Whenever you need a friend.
('Cause I get you and you get me.)
Whenever you need a friend.
I can't pretend
there's anywhere else I'd rather be,
'Cause you've got a friend in me.

'Cause I get you and you get me.
(Yeah, I get you!)
'Cause I get you and you get me.
(I think you get me too.)
We're a special kind of family.

Let's sing it!

You've got a friend,
('Cause I get you and you get me.)
Whenever you need a friend.
You've got a friend,
('Cause I get you and you get me.)
Whenever you need a friend.
You've got a friend,
('Cause I get you and you get me.)
Whenever you need a friend.
We're a special kind of family.

You've got a friend,
('Cause I get you and you get me.)
Whenever you need a friend.
You've got a friend,
('Cause I get you and you get me.)
Whenever you need a friend.
You've got a friend,
('Cause I get you and you get me.)
Whenever you need a friend.
We're a special kind of family.

You've got a friend,
('Cause I get you and you get me.)
Whenever you need a friend.
You've got a friend,
('Cause I get you and you get me.)
Whenever you need a friend.
You've got a friend,
('Cause I get you and you get me.)
Whenever you need a friend.
We're a special kind of family.

Tiny Tapas

Words and Music by
Michael Wiskar

Olé, Olé, Olé! (Welcome To Spain)

Words and Music by
Michael Wiskar

C E7 Ami
41
stay And dance the night a - way. O - lé, o - lé, o - lé!
Dmi Ami
47
Four, Eight, Twelve, Six - teen You, my friends, are pret - ty nif - ty. Twen - ty, Twen - ty - four,
Dmi C
52
Twen - ty - eight, Thir - ty - two Now my friends here's what we're going to do. Five, Ten, Fif - teen, Twen - ty
G E7
57
You are in a place of plen - ty. Twen - ty - five, Thir - ty, Thir - ty - five, For - ty Now it's time to
Ami Dmi
62
start the par - ty! O - lé, o - lé, o - lé! Wel - come to spain. Let the mu - sic
C E7 Ami
67
play 'Til the break of day. Oh oh oh oh O - lé, o - lé, o - lé!
Dmi C
72
Wel - come to Spain. We'd love it if you'd stay And dance the night a -
E7 Ami
77
way. O - lé, O - lé, o - lé. O - lé!

Break It Down Into Parts

Words and Music by
Michael Wiskar

Ami7
22
Here's some - thing you've got to know. You've got to break it down, break it down, you've got to
C/B♭ C
24
break it down in - to parts. If your trip is real - ly long, just lis - ten to this sil - ly song and
Ami7 C/B♭ C
27
break it down, break it down, you've got to break it down in - to parts. Third part. What d'ya see?
Ami B♭ C Ami B♭
30
Two stone lions three quar-ters of the way now. Fourth part. What d'ya see? You're at the park yeah, you made it all the way now.
C C Ami7
33
(Yeah) When you've got so far to go, here's some-thing you've got to know. You've got to break it down, break it down, you've got to
C/B♭ C
37
break it down in - to parts. If your trip is real - ly long just lis - ten to this sil - ly song and
Ami7 C/B♭ C
40
break it down, break it down, you've got to break it down in - to parts. Break it down in - to parts.

♩ = 120 # Almost Home (The Perimeter Song)

Words and Music by
Michael Wiskar

D/A Bmi7
34
We go trav - 'ling 'round the per - im - i - ter.
D Bmi7 D
39
Up one side a - cross the o - ther. 'Round and down the per - im - i - ter. One more side and
A D/A
44
then we're al - most home Then we're al - most
A D/A A
49
home And Then we're al - most home
D/A A D/A
55
And Then we're al - most home Then we're al - most
A D/A Ama7 D/A
61
home.
Ama7 D/A Ama7 D6
69

BEST FRIENDS
(I GET YOU AND YOU GET ME)

WORDS AND MUSIC BY
MICHAEL WISKAR

D/A F♯mi E A D/A
34
thing to do we're ne-ver feel - ing bored. We've had a thou - sand ad-ven - tures and we'll have
F♯mi E A/C♯ D F♯mi E
39
a thou - sand more. You make me laugh so hard I feel like I could cry.
A/C♯ D F♯mi E
44
And we pick each o - ther up when we're to-ge-ther, we don't ev - en have to try. You've got a
D.S. AL CODA
CODA
D/E A Emi7 A
49
you've got a friend in me. Yeah 'Cause I get you and you get me. 'Cause
Emi7 A Emi7 DMA9
9fr.
53
I get you and you get me. We're a spe-cial kind of fa - mi - ly.
F♯mi E G F♯mi
57
And ev-en if there's trou - ble brew - ing we al-ways know just what we're do - ing, 'Cause I've
B D/E E7
62
al-ways got your back and you've got mine. And ev - 'ry-thing's fine. You've got a
D.S. AL CODA

CODA
D/E
A
D/G
A
D/G
I get you and you get me. 'Cause I get you and you get me. need a friend. I can't pre-
When-ev-er you need a friend. When-ev-er you
F♯mi
A/C♯
Bmi
D/E
A
tend there's an-y-where else I'd ra - ther be. 'Cause you've got a friend in me.
'Cause
Emi7
A
Emi7
A
Emi7
I get you and you get me. 'Cause I get you and you get me. We're a spe-cial kind of fa -
Yeah, I get you. I think you get me too.
DMA9
A
Emi7
A
mi - ly. 'Cause I get you and you get me. 'Cause I get you and you
Let's sing it!
You've got a friend, when - ev - er you need a friend. You've got a friend, when - ev - er you
Emi7
A
Emi7
Emi/D
1, 2.
DMA9
get me. 'Cause I get you and you get me. We're a spe-cial kind of fa - mi-ly. 'Cause
need a friend. You've got a friend, when - ev - er you need a friend. We're a spe-cial kind of fa - mi-ly. You've got a
3.
DMA9
A
mi - ly.